Kyung Chun Kim
Suresh Baral

Solar Organic Rankine Cycle Power System for Developing Countries

Kyung Chun Kim
Suresh Baral

Solar Organic Rankine Cycle Power System for Developing Countries

An Appropriate Technology

LAP LAMBERT Academic Publishing

Cover image: www.ingimage.com

Publisher:
LAP LAMBERT Academic Publishing
is a trademark of
Dodo Books Indian Ocean Ltd. and OmniScriptum S.R.L publishing group

120 High Road, East Finchley, London, N2 9ED, United Kingdom
Str. Armeneasca 28/1, office 1, Chisinau MD-2012, Republic of Moldova, Europe
Managing Directors: Ieva Konstantinova, Victoria Ursu
info@omniscriptum.com

Printed at: see last page
ISBN: 978-3-659-76734-0

Solar Organic Rankine Cycle Power System for Developing Countries: An Appropriate Technology

Suresh Baral

Kyung Chun Kim

School of Mechanical Engineering

Pusan National University, Busan, Republic of Korea

2015

Preface

The primary objective of this book is to recommend the solar ORC system as an appropriate technology for electricity generation in remote areas of developing countries. For this concept, the solar ORC system prototype has been built, installed and tested in the laboratory. From the experimental results, it is suggested that the system is robust in its configuration and is suitable for installing in remotes places. Besides, the economic analysis was carried that shows that it is feasible with some financial subsidies. The book contains four chapters from the fundamental concept to the economics aspect.

Chapter 1 discusses the importance of the solar ORC system in developing countries, and the thermodynamic theories to provide background for the investigated the system. This includes the fundamental laws of thermodynamics with their governing equations. The energy, mass and exergy balance equations are discussed to support the theoretical and experimental results.

Chapter 2 covers performances as well as thermodynamic and environmental properties of different working fluids. Besides, these performance parameters are comparatively evaluated for the use of solar ORC system. The comparison parameters are exergy and thermal efficiencies, pressure ratio, heat input, area of collector, safety, ozone depletion potential and global warming potential.

Chapter 3 discusses for experimental setups. It was constructed in the laboratory for testing the system's performance. The results of the tests were carried out with R245fa as a working fluid and are discussed in this chapter. Furthermore, the simulation for the power output from solar ORC has been incorporated.

The book is targeted to the solar ORC developers, manufacturers and energy planners, so it is very crucial to address the issue of total installation cost of solar ORC system. For this different financial condition is developed in order to analyze the cost of electricity generation from the system. Based on the experimental analysis of the system, the thermoeconomic analysis is carried out and is discussed in this **Chapter 4** along with other economic parameters such as payback period, internal rate of return and sensitivity analysis. Finally in this very chapter, the concept and economic viability assessment was carried out for the solar ORC water pumping system. Currently, the cost of water pumping is high but in long run, it could be one of the multi-purpose systems that can have electricity, hot water production, heating, cooling and pumping of water from the same unit.

Contents

List of figures

List of tables

List of symbols

a	area, m^2
c	constant specific heat capacity, kJ /kg K
F_R	collector heat –removal factor
S	solar insolation, W/m^2
h	enthalpy, kJ /kg
$\dot{m}$	mass flow rate, kg/s
N	total number
$\dot{Q}$	heat transfer, kW
s	entropy, kJ/ kg K
T	temperature, °C
t	time, s
U_L	overall heat coefficient, W/m^2K
V	volume of hot water, L
$\dot{W}$	work transfer, kW
Q	discharge, m^3/s
P	Power, kW
H	head, m

Greek symbols

η efficiency

α absorptivity

ρ density

τ transmissivity

Subscripts and superscripts

b Beam

c Collector

C Cost

cog Co-generation

con Condenser

d Diffuse

el Electrical

eva Evaporator

exp Expander

hw Hot water

i Inlet

in Inlet

n	Economic life
out	Outlet
p	Pump
sol	Solar
spc	Solar power cycle
t	Turbine
th	Thermal
u	Useful
wf	Working fluid

Acronyms

AB	Annual benefit
AC	Annual cost
BC	Benefit-Cost
GWP	Global warming potential
IC	Investment cost
IR	Interest rate
IRR	Internal rate of return
LCOE	Levelized cost of electricity

NPV	Net present value
ODP	Ozone depletion potential
ORC	Organic Rankine Cycle
PBP	Payback period
PV	Photovoltaic
SORC	Solar organic Rankine cycle

Chapter One: Introduction

1.1 Importance of solar organic Rankine cycle system technology

Appropriate Technology is an approach to technological development, which is characterized by original and sound engineering, which differentiates the social, environmental, political, economic, and technical aspects of a proposed technological solution to a problem facing a society. Usually, appropriate technologies are smaller-scale technologies that are ecologically and socially benign, affordable, and powered by renewable energy sources. Areas of interest include energy conversion systems, waste and water management, community and shelter design, technology assessment, small-scale production systems, and technology transfer [1].

Energy poverty refers to a lack of access to modern energy services, and the well-being of a large proportion of people in developing countries is affected depressingly by the very low consumption of clean energy. Modern energy services are crucial to human well-being and to a country's economic development. Access to modern energy is essential for the provision of clean water, sanitation and healthcare and for the provision of reliable and efficient lighting, heating, cooking, mechanical power, and transport and telecommunications services. It is an alarming fact that today billions of people lack access to the most basic energy services: as World Energy Outlook 2014 shows nearly 1.3 billion people are without access to electricity and 2.7 billion people rely on the traditional use of biomass for cooking, which causes harmful indoor air pollution. These people are mainly in either developing Asia or sub-Saharan Africa, and in rural areas [2].

In addition, 85% of the people without electricity live in rural areas in developing countries and the majority of these people are found in Sub-Saharan and South Asia [2]. Therefore the rural poor without access to electricity use kerosene-based lighting, particularly the open fire, simple wick kerosene candle. This is the cheapest lighting option but can provide only dim light, creates indoor air pollution, poses a serious fire risk, and is a significant contributor to respiratory diseases that kill more than 1.5 million people every year [2]. Therefore, there is urgent need to solve this energy poverty by providing clean, safe and reliable sources of energy for lighting houses. The average household of four people uses approximately 300 kWh per year for lighting alone [3]. The estimated amount of electricity required for people to read at night, pump a minimal amount of drinking water and listen to radio broadcasts is only 50 kWh per person per year [4]. In many rural areas of developing countries, electric grid connection is economically not feasible or may take decades to arrive. Currently, there is a wide range of viable and cost-competitive renewable energy alternatives that can be powered by solar energy. Among different types of appropriate technologies, solar organic Rankine cycle system could be one of the multi-purpose technologies for rural communities for producing electricity and hot water in the rural homes.

In recent years, organic Rankine cycle has become a field of intense research and appears as a promising technology for conversion of heat into useful work or electricity. Organic Rankine cycle (ORC) can be applied in various applications such as in geothermal [5-8], biomass [9-12] and waste heat recovery [13-17]. The other application includes solar energy utilization. The use of solar energy for generating electricity on a micro- and small-scale using organic liquids as the working fluid in the solar Rankine cycle system is important and is expected to be popular in developing countries.

Some articles have discussed adapting the small scale solar ORC system design to electricity generation in rural poor villages [18–23]. Reducing the specific investment cost as well as operation and maintenance cost for a very small scale solar ORC system is crucial. Therefore, the group of researchers at Massachusetts Institute of Technology (MIT) in collaboration with the University of Liege and the non-governmental organization, Solar Turbine Group (STG), developed and implemented a small-scale solar ORC in the rural African country of Lesotho, with a power output of 3 kW and obtained an overall electrical efficiency of 7% to 8% and figure 1.1 shows setup [24].

Figure 1.1 Solar ORC field trails in Lesotho (Photo courtesy by Solar Turbine Group, USA)

1.2 Working principle of solar organic Rankine cycle (SORC) system

The working principle of solar organic Rankine cycle (SORC) system is described briefly. Solar radiation is incident to an evacuated tubular type solar collector. Cold water is passed through it and the heated water is stored in a storage tank. The hot water from the tank is pumped through the pump and passed into the plate type heat exchanger (evaporator). The organic working fluid, refrigerant R245fa, vaporizes when passed into the evaporator, which drives the magnetically coupled scroll expander to produce mechanical power. The working fluid changes its phase from a vapor to liquid state when the cooling water from the tap is passed into the condenser. Lastly in its liquid form, the working fluid runs into the refrigerant tank to complete its cycle. The mechanical power produced by the ORC unit is then coupled to an electrical generator for electricity generation. Figure 1.2 shows the schematic diagram for solar ORC system.

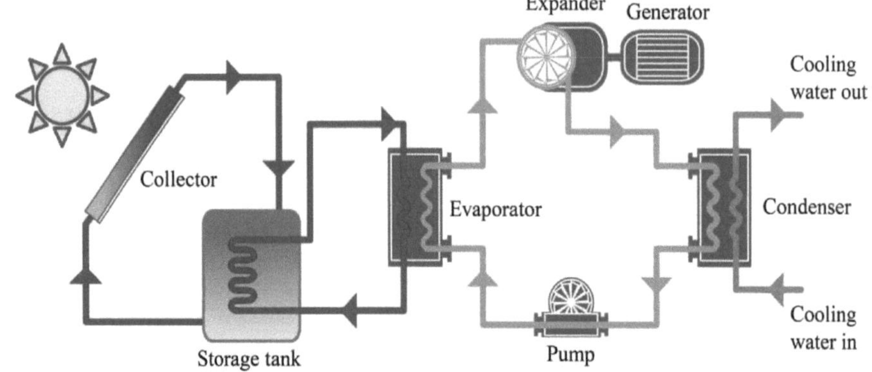

Figure 1.2 Schematic diagram of a solar ORC system

The energy balance is applied to each of the system components based on the first law of thermodynamics. The general energy balance equation in steady state for any components can be written as follows [25]:

$$\sum \dot{m}_{in} = \sum \dot{m}_{out} \tag{1}$$

$$\dot{Q} - \dot{W} + \sum \dot{m}_{in} h_{in} - \sum \dot{m}_{out} h_{out} = 0 \tag{2}$$

where subscripts in and out represent the inlet and outlet, respectively; $\dot{m}$ and h represents the mass flow rate and specific enthalpy of the streams of the system working fluid, respectively, and $\dot{Q}$ and $\dot{W}$ represent the heat transfer crossing the component boundaries, respectively. Figure 1.3 shows the T-s diagram for the ORC process.

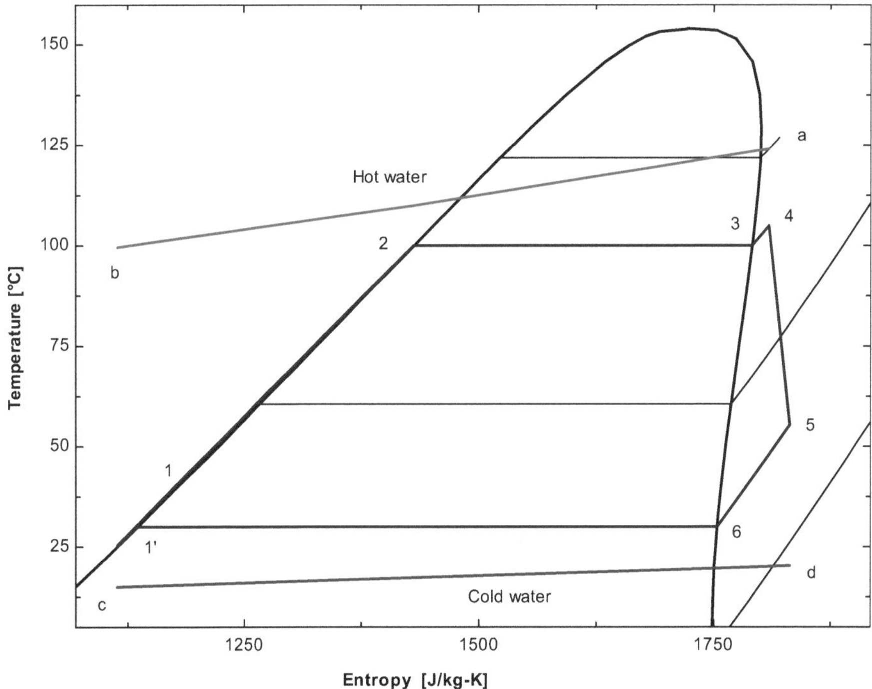

Figure 1.3 T-s diagram for ORC process

The heat energy required for heating working fluid for the ORC system can be expressed by the following equation:

Evaporator, $\dot{Q}_{eva} = \dot{m}_f \left(h_2 - h_4\right)$ (3)

The expansion device for work done can be calculated by using the following expression:

Expander, $\dot{W}_{\exp} = \dot{m}_f \left(h_6 - h_4\right)$ (4)

The heat rejected to the environment can be expressed by:

Condenser, $\dot{Q}_{con} = \dot{m}_f \left(h_6 - h_{1'}\right)$ (5)

The pumping of working fluid to the heat exchanger can be expressed as

Pump, $\dot{W}_p = \dfrac{\dot{m}_f v_{1'} \left(P_{1'} - P_1\right)}{\eta_p}$ (6)

Here *h* , *v* , *m, and P* represent enthalpy, volumetric flow , working fluid mass flow rate and pressure at different states. The efficiency of the working fluid feed pump is denoted by η_p.

1.3 Solar energy collecting system

1.3.1 Flate plate collector

In solar energy application, flate type solar collectors are used to collect the solar radiation from the sun. When these solar radiations falls on the collector glass cover, it is absorbed by absorber plate which in turns carried away by water in the tubes attached to the collector plate. The hourly radiation falling on a tilted surface is given by the following expression [26]:

$I_T = I_b R_b + I_d R_d + \left(I_b + I_d\right) R_r$ (7)

Where, I_b and I_d are the hourly beam and diffuse radiation that the collector receives, and R_b, R_d, R_r are defined as tilt factors for different kinds of radiation, whose values are given by :

$$R_b = \frac{\cos(\theta)}{\cos(\theta_z)} = \frac{\sin\delta\sin(\phi-\beta)+\cos\delta\cos\omega(\phi-\beta)}{\sin\phi\sin\delta+\cos\phi\cos\delta\cos\omega} \tag{8}$$

for beam radiation, where $\delta$is the declination angle .

$$\delta = 23.45\sin\left[\frac{360}{365}(284+n)\right] \tag{9}$$

$$R_d = \frac{1+\cos\beta}{2} \tag{10}$$

for diffuse radiation, and

$$R_r = \rho\left(\frac{1-\cos\beta}{2}\right) \tag{11}$$

for reflected radiation falling on the surface of the plate.

The incident solar flux absorbed in the absorber plate is given by

$$S = I_b R_b(\tau\alpha)_b + [I_d R_d + (I_b + I_d)R_r](\tau\alpha)_d \tag{12}$$

where $(\tau\alpha)_b$ and $(\tau\alpha)_d$ represent the transmissivity-absorptivity product for beam and diffuse radiation falling on the collector respectively.

The useful heat gain rate for the collector is given by the following expression:

$$Q_u = F_R . A_c . [S - U_L . (T_i - T_a)] \tag{13}$$

where, F_R , U_L, A_c, T_i and T_a are collector heat –removal factor, total loss coefficient, thermal-absorption area of the collector, water inlet temperature and ambient temperature respectively.

1.3.2 Vacuum type collector

The solar collector efficiency can be calculated as function of solar incident, mean temperature of collector and ambient air temperature and can be given by following expression [27]:

$$\eta_c = C_o - \frac{C_1(T_i - T_a)}{I_{sol}} - \frac{C_2(T_i - T_a)^2}{I_{sol}} \tag{14}$$

where, S is the solar flux absorbed in the absorber plate , C_o, C_1, C_2 are coefficient constants for collector. If it is assumed that solar collector cover is glass plate with very small order of thickness and that solar heat flux variations have a 24 hours period, the solar collector can be assumed in a state of steady. So, in this case, the useful thermal energy absorbed by the solar-collector fluid is given by following expression:

$$\dot{Q}_u = \eta_c . S . A_c \tag{15}$$

where, A_c is the area of solar collector.

The net electrical output power:

$$\dot{W}_{net} = \dot{W}_{\exp} - \dot{W}_p \tag{16}$$

The Solar ORC cycle efficiency:

$$\eta_{ORC} = \frac{\dot{W}_{net}}{\dot{Q}_u} \tag{17}$$

1.4 Advantages of solar ORC system

Even though the cost of ORC system is high it has following benefits that may make this system more feasible for the rural application:

1. Easy to install ORC configuration: Since rural areas lack electricity, it is very easy to install the ORC units within short period of time.
2. Greater equipment longevity: The mechanical stresses on the ORC components are lower due to low pressure and low scroll expander rotational speed (maximum 3600 RPM), the ORC unit can have longer life. Also the parts are easily available in the local market.
3. Potential use for co-generation/tri-generation applications: The same ORC unit can be used for co-generation applications such as electricity, space heating and cooling and domestic hot water production for rural people living in different climatic conditions.
4. Environmental cost benefits: If environmental costs are considered, renewable energy sources such as solar ORC system is far better than diesel generators, kerosene base lighting. This helps rural people live healthier and uplift the living standard.
5. No grid extension cost: Since the ORC system is off-grid system to produce electricity; there is no need of grid- extension. This makes rural electrification more feasible by ORC technology.

1.5 Expected socio-economic benefits by the solar ORC system

The socio-economic benefits of solar ORC system are listed below:

Economic	Job creation: The increased number of jobs directly or indirectly created by the solar ORC technology (staff to operate and maintain the ORC facilities, increased economic activity by small home businesses enterprises and productive users). Household income: There is an increase in household income after the provision of electricity. Cut-off in household expenditures: No need to purchase kerosene fuel or other fuels for the lighting of houses. Economic development: By the solar ORC technology it can improve the overall income growth, income per capita, poverty alleviation, *etc.* thereby uplifting the living standards of the people.
Educational Benefits	Improve quality of teaching and learning process in schools through the provision of electricity-dependent equipment, such as computers, printer, and overhead projector. Increase the study time for children at home during night time by lighting. Improve access to communication devices, such as radio/TV and mobile phones.
Social welfare	Health benefits: Improvements to the community health post, clinics (cooling, lighting); better health due to cleaner air as households reduce the use of polluting fuels for lighting and cooking (indoor-lighting); improve health knowledge through increased access to information on radio/TV.

	Social benefits: Increased time spent on community activities for the development of community rather than collecting firewood for lighting.
Environmental	Global environmental benefits: Decrease in greenhouse gases (GHG) emissions and the utilization of clean and green renewable resources.

References

1. http://appropriatetec.appstate.edu/research-reports-articles(Accessed :2015/04/20)
2. IEA (2010). Comparative Study on Rural Electrification Policies in Emerging Economies; International Energy Agency (IEA): Paris, France.
3. Zahnd, A., & Kimber, H. M. (2009). Benefits from a renewable energy village electrification system. Renewable Energy, 34(2), 362-368.
4. Ramage, J (1997). Energy a Guidebook; Oxford University Press: Oxford, UK.
5. Bronicki, L. Y. (1988). Electrical power from moderated temperature geothermal sources with modular mini-power plants. Geothermics, 17(1), 83-92.
6. Bronicki LY. (2007). Organic Rankine Cycles in Geothermal Power Plants – 25 years of ORMAT experience, GRC Annual Meeting Reno, NV.
7. Brasz JJ, Biederman BP, Holdmann G. (2005). Power Production from a Moderate Temperature – Geothermal Resource GRC Annual Meeting, Reno, NV, USA.
8. Aneke, M., Agnew, B., & Underwood, C. (2011). Performance analysis of the Chena binary geothermal power plant. Applied Thermal Engineering, 31(10), 1825-1832.
9. Borsukiewicz-Gozdur, A., Wiśniewski, S., Mocarski, S., & Bańkowski, M. (2014). ORC power plant for electricity production from forest and agriculture biomass. Energy Conversion and Management, 87, 1180-1185.
10. Noussan, M., Abdin, G. C., Poggio, A., & Roberto, R. (2014). Biomass-fired CHP and heat storage system simulations in existing district heating systems.Applied Thermal Engineering, 71(2), 729-735.
11. Algieri, A., & Morrone, P. (2014). Energetic analysis of biomass-fired ORC systems for micro-scale combined heat and power (CHP) generation. A possible application to the Italian residential sector. Applied Thermal Engineering, 71(2), 751-759.

12.Jradi, M., & Riffat, S. (2014). Modelling and testing of a hybrid solar-biomass ORC-based micro-CHP system. International Journal of Energy Research,38(8), 1039-1052.

13.Soffiato, M., Frangopoulos, C. A., Manente, G., Rech, S., & Lazzaretto, A. (2015). Design optimization of ORC systems for waste heat recovery on board a LNG carrier. Energy Conversion and Management, 92, 523-534.

14.Song, S., Zhang, H., Lou, Z., Yang, F., Yang, K., Wang, H., & Yao, B. (2015). Performance analysis of exhaust waste heat recovery system for stationary CNG engine based on organic Rankine cycle. Applied Thermal Engineering, 76, 301-309.

15.Imran, M., Park, B. S., Kim, H. J., Lee, D. H., & Usman, M. (2015). Economic assessment of greenhouse gas reduction through low-grade waste heat recovery using organic Rankine cycle (ORC). Journal of Mechanical Science and Technology, 29(2), 835-843.

16.Zare, V., & Mahmoudi, S. M. S. (2015). A thermodynamic comparison between organic Rankine and Kalina cycles for waste heat recovery from the Gas Turbine-Modular Helium Reactor. Energy, 79, 398-406.

17.Zhang, C., Shu, G., Tian, H., Wei, H., & Liang, X. (2015). Comparative study of alternative ORC-based combined power systems to exploit high temperature waste heat. Energy Conversion and Management, 89, 541-554.

18.Baral, S., Kim, D., Yun, E., & Kim, K. C. (2015). Energy, exergy and performance analysis of small-scale organic Rankine cycle systems for electrical power generation applicable in rural areas of developing countries. Energies, 8(2), 684-713.

19.Baral, S., Kim, D., Yun, E., & Kim, K. C. (2015). Experimental and Thermoeconomic Analysis of Small-Scale Solar Organic Rankine Cycle (SORC) System. Entropy, 17(4), 2039-2061.

20. McMahan, A. C. (2006). Design & optimization of organic Rankine cycle solar-thermal power plants (Master's dissertation, University of Wisconsin--Madison).
21. Georges, E., Declaye, S., Dumont, O., Quoilin, S., & Lemort, V. (2013). Design of a small-scale organic Rankine cycle engine used in a solar power plant. International Journal of Low-Carbon Technologies, ctt030.
22. Quoilin, S., Orosz, M., Hemond, H., & Lemort, V. (2011). Performance and design optimization of a low-cost solar organic Rankine cycle for remote power generation. Solar Energy, 85(5), 955-966.
23. Kane, M., Larrain, D., Favrat, D., & Allani, Y. (2003). Small hybrid solar power system. Energy, 28(14), 1427-1443.
24. Orosz, M. S., Quoilin, S., & Hemond, H. (2013). Technologies for heating, cooling and powering rural health facilities in sub-Saharan Africa. Proceedings of the Institution of Mechanical Engineers, Part A: Journal of Power and Energy, 227(7), 717-726.
25. Dincer, I., & Rosen, M. A. (2012). Exergy: energy, environment and sustainable development. Newnes.
26. Wang, M., Wang, J., Zhao, Y., Zhao, P., & Dai, Y. (2013). Thermodynamic analysis and optimization of a solar-driven regenerative organic Rankine cycle (ORC) based on flat-plate solar collectors. Applied Thermal Engineering, 50(1), 816-825.
27. Freeman, J., Hellgardt, K., & Markides, C. N. (2015). An assessment of solar-powered organic Rankine cycle systems for combined heating and power in UK domestic applications. Applied Energy, 138, 605-620.

Chapter Two: Selection of working fluids for solar ORC system

Fluid selection is one of the most important contributors to overall solar ORC cycle performance. Fluid thermodynamics, material compatibility, flammability, toxicity and other properties must be considered with respect to the system's needs. Desirable characteristics of the working fluids include appropriate boiling point temperature, low latent heat, high critical temperature and pressure, suitable specific volume, low density and surface tension, high thermal conductivity, high thermal stability, non-corrosive, non-toxic, and compatibility with engine materials [1-9]. It is necessary to find the working fluid as a potential candidate among different types. For low- temperature solar ORC such as the one that uses flat plat collector, the working fluids critical temperature ranges from 95 °C to 135°C whereas for medium- temperature solar ORC that uses vacuum type and parabolic type of solar collectors, critical temperature ranges from 150 °C to 240 °C. Table 2.1 shows the potential candidates for the solar ORC system based on the thermo-physical characteristics of the fluids. The simulation is carried out in order to recommend the appropriate candidate for the solar ORC system. For the simulation, the desired thermodynamic properties of the ORC working fluid at different state points should be known. The thermodynamic properties of an ORC working fluid can be described best by the energy equation and calculated using software known as an Energy Equation Solver (EES)[10] Therefore, all the thermodynamics properties of ORC working fluids in this study were obtained using software.

Table 2.1 Properties of various working fluids for solar ORC system

Working Fluid		Molecular mass (kg/kmol)	Critical Temperature (°C)	Critical Pressure (MPa)	ASHRAE safety group	Atmospheric life time(yr)	ODP	GWP (100 yr)
1	R22	86.5	96.14	4.9	A1	12.1	0.055	1500
2	R290	44.1	96.68	4.2	A3	0.041	0	20
3	R134a	102.03	101	4.059	A1	14	0	1430
4	R227ea	170	101.7	2.9	A1	34.2	0	3220
5	R500	99.3	105.6	4.17	A1	74	0.74	6010
6	R12	120.91	112	4.11	A1	100	0	10890
7	RC318	200.03	115.2	2.79	A1	3200	0	10250
8	R717	17.03	132.3	11.33	B2	0.01	0	1
9	R600a	58.12	134.7	3.64	A3	12	0	n.a
10	R245fa	134.05	154.2	3.64	B1	8.8	0	820
11	R123	153	183.7	3.668	B1	1.3	0.02	77
12	R11	137.4	198	4.4	A1	50	1	5800
13	R141b	116.95	204.2	4.25	n.a	9.3	0.12	725
14	Methanol	32.04	240.2	8.1	n.a	n.a	n.a	n.a
15	Ethanol	46.07	240.8	6.14	n.a	n.a	n.a	n.a

2.1 Assumption and simulation environment for ORC

For simplicity, several reasonable assumptions were implemented by thermodynamic modeling and shown in Table 2.2.

The reasons for assumptions are justified by following points:

1. The expander power output is fixed at 1 kW because the power could be more than 1 kW according to the specification. Also the specification of scroll expander has maximum of 175 °C inlet source temperature. The proposed modelling is only for 85 °C and 120 °C. So the power output is assumed to be 1 kW.
2. Due to different geographic location and conditions, the solar insolation/irradiance is different. It is assumed that low solar irradiance location or during the winter season, the hot water could be only 85 °C. On other hand, if the solar irradiance is high and during

the summer season, the solar collector temperature could be reached up to 120 °C with simple flat or vacuum type collector.

3. From the thermodynamic modelling, it is also proposed to find the quantity of hot water produced in a day. Normally the hot water having the temperature of 45 °C can be useful for utilizing in domestic purposes. So the condensation temperature is fixed at 45 °C.
4. There are several literatures that discuss the efficiencies of scroll expander, pump and solar collector to be within 70 % [11]. So in this thermodynamic modelling, it is assumed to be the same.
5. In the ORC system, the pressure losses are very negligible so it is neglected in this study [11].

Table 2.2 Assumptions for thermodynamic modeling

S.N	Assumptions	Value
1	Expander power output	1 kW
2	Low/medium solar ORC temperature	85°C/ 120 °C
3	Condensation temperature	45 °C
4	Expander efficiency	70 %
5	Pump efficiency	70 %
6	Solar collector efficiency	70 %
7	Pressure drop in ORC	Negligible

2.2 Results and discussion

In total, 15 pure organic working fluids were selected as potential candidates. Hot water is provided by solar collectors. The condenser is cooled by tap water of temperature 25 °C. In this study, it is well-thought-out that vapor at the turbine inlet is saturated. The re-heater was not introduced because it could be expensive system. The system performance results are displaced in Table 2.3 for a design of 1kW power output. The study of the ORC system is located in Busan, South Korea whose latitude and longitude are 35.15° N, 129.06° E respectively [12].

Table 2.3 Comparison of the performances of different working fluids for a 1 kW power output

Working Fluid	P_{max} (kPa)	P_{min} (kPa)	PR	Q_{in} (kW)	$\dot{m}_{wf}$ (kg/s)	η_{exg} (%)	η_{th} (%)	η_{SPC} (%)	$\dot{V}$ (L/s)	$A_c(m^2)$
R22	4037	1730	2.33	17.99	0.12	31.04	5.55	3.89	0.52	32.13
R290	3437	1534	2.24	18.57	0.06	30.08	5.38	3.77	0.63	33.16
R134a	2928	1161	2.52	17.41	0.1	32.08	5.74	4.02	0.59	31.09
R227ea	2075	638	3.25	15.04	0.12	37.14	6.65	4.65	0.58	26.85
R500	3015	1282	2.35	19.07	0.13	29.29	5.24	3.67	0.73	34.05
R12	2536	1084	2.34	17.88	0.14	31.23	5.6	3.91	0.82	31.93
RC318	1500	289	5.19	11.59	0.08	48.19	8.63	6.04	0.54	20.69
R717	4609	1782	2.58	14.74	0.02	37.88	6.78	4.75	0.37	26.32
R600a	1462	469	3.11	13.25	0.04	42.13	7.54	5.28	0.9	23.67
R245fa	1929	181	10.66	8.51	0.04	46.82	11.76	8.23	0.3	15.18
R123	1201	111	10.82	7.95	0.04	50.03	12.58	8.8	0.5	14.2
R11	1228	202	6.08	9.09	0.04	43.76	11	7.69	0.68	16.24
R141b	1031	126	8.18	8.49	0.03	46.84	11.77	8.23	0.68	15.17
Methanol	631	44	14.34	8.18	0.006	48.62	12.22	8.55	0.95	14.62
Ethanol	429	23	18.65	8.37	0.006	47.55	11.95	8.36	1.24	14.95

2.2.1 Cycle efficiencies

The ORC efficiency is the most important index used to evaluate performance of the system. Nine working fluids emerged as potential candidates for low-temperature and six working fluids for medium-temperature solar ORC .The system exergy efficiency ranges from 29.9 % 50.3%. It is seen that maximum exergy efficiency is found with RC318 and R123 as shown in Table 2.3. The thermal efficiency and solar power cycle efficiency are in the ranges between 5.38 % to 12.58 % and 3.67 % to 8.8% respectively. The system exergy efficiency of ORC varies with turbine inlet temperature as shown in figures.2.1 and 2.2. It is due to the working fluids characteristics when subjected to various temperatures.

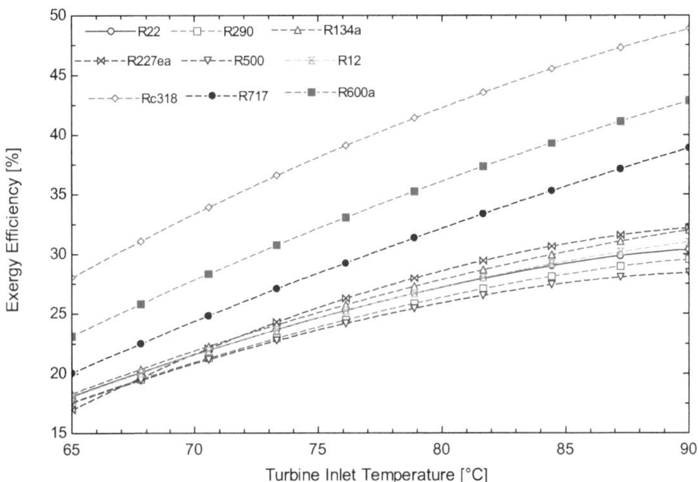

Figure 2.1 Variation of the exergy efficiency for low-temperature solar ORC working fluids with the turbine inlet temperature at the condensing temperature of 45°C

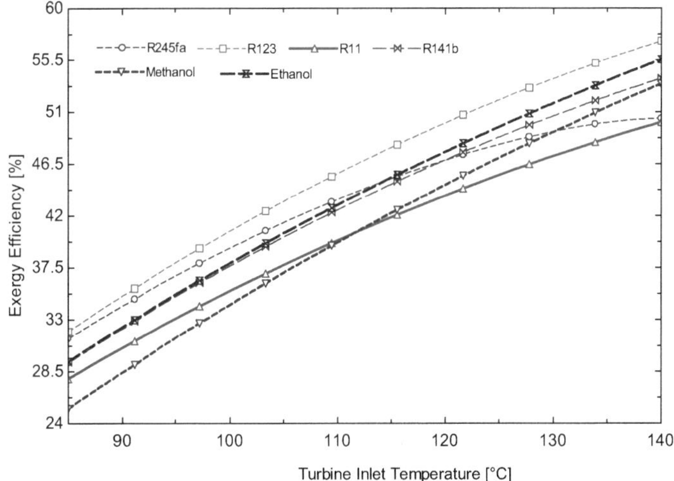

Figure 2.2 Variation of the exergy efficiency for medium-temperature solar ORC working fluids with the turbine inlet temperature at a condensing temperature of 45°

2.2.2 Cycle pressure ratio

The cycle pressure ratio determines the type of expander size and number in ORC system. Higher pressure ratio accounts for installation of the number of expanders arranging in series or parallel in the ORC system. High pressure ratio also increases the efficiency of the system. From the Table 2.3, R227ea, RC318 and R600a, R123, methanol and ethanol have low condensing pressure so they show higher pressure ratio. High turbine inlet temperature yield high pressure ratio which can be seen from the figures 2.3 and 2.4 when temperature varies from 85 °C to 120 °C.

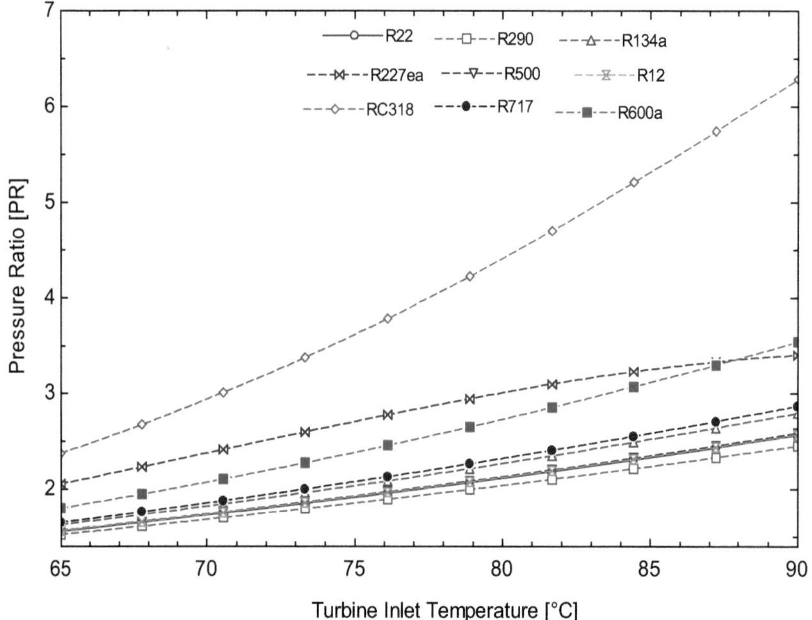

Figure 2.3 Variation of pressure ratio for low-temperature solar ORC working fluids with turbine inlet temperature at condensing temperature of 45°C

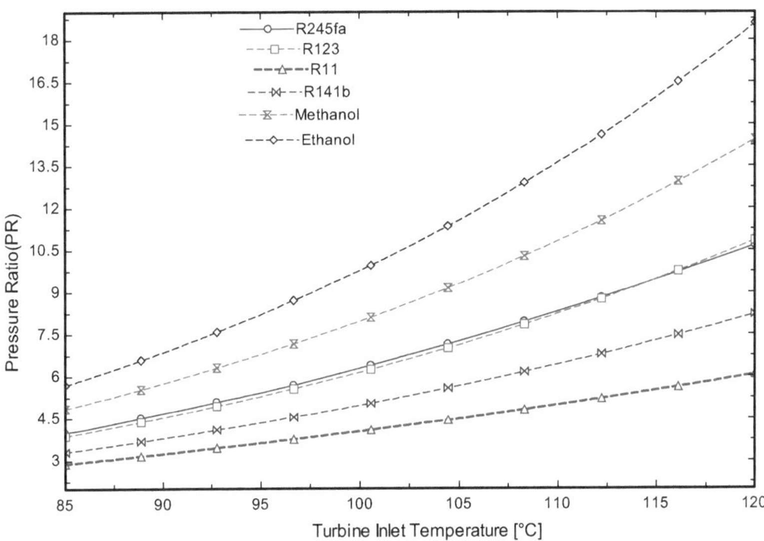

Figure 2.4 Variation of pressure ratio for medium-temperature solar ORC working fluids with turbine inlet temperature at condensing temperature of 45°C

2.2.3 Flow rates

For economic aspects, the turbine outlet volume flow rate plays an important role because it determines the size and cost. As shown in Table 2.3, R12, R500, R600a, methanol and ethanol exhibited a high volume flow rate. Fluids with a low volume flow rate are preferable for decreasing the pump size and friction loss in a pipe system. Among these are R245fa, R717, R123, R22, RC318, R227ea, and R134a. In general, the turbine outlet volume flow rate is inversely proportional to the turbine inlet temperature, which can be illustrated in figures 2.5 and 2.6. For the low-temperature solar ORC, R717 shows the best performance followed by R22, RC318 and R134a. For the medium-temperature solar ORC, R245fa shows the lowest turbine inlet volume flow rate followed by R123 and R11.

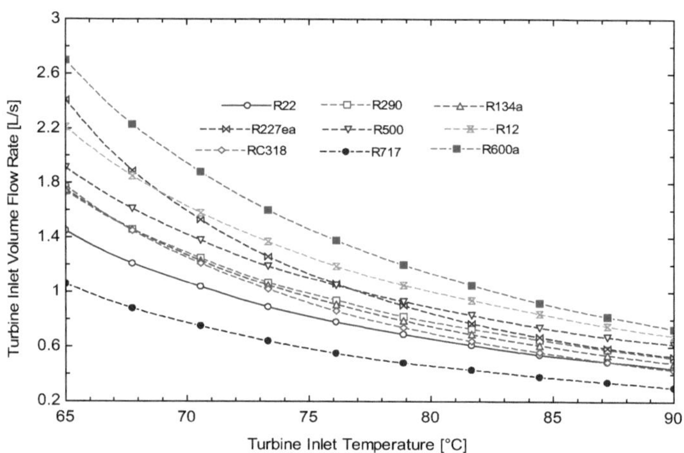

Figure 2.5 Variation of turbine outlet volume flow rate for low-temperature solar ORC working fluids with a turbine inlet temperature at a condensing temperature of 45°C

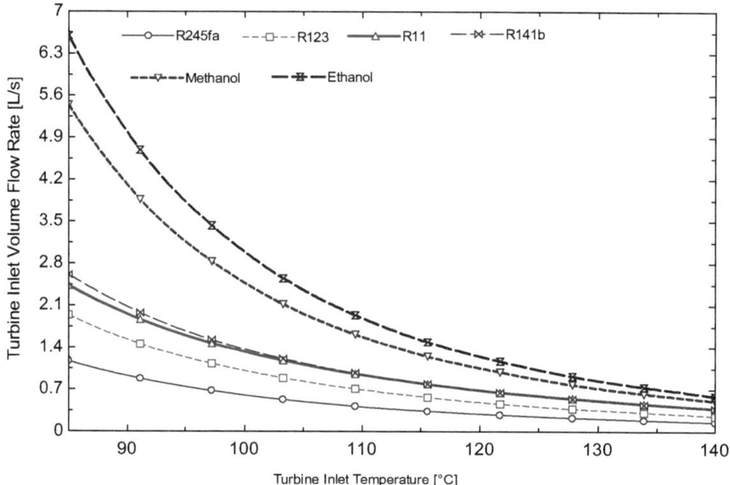

Figure 2.6 Variation of turbine outlet volume flow rate for medium-temperature solar ORC working fluids with a turbine inlet temperature at a condensing temperature of 45°C

2.2.4 Area of solar collector

The heat input to the system is significant in a solar ORC that determines the size of the collector and establishes a major part of the system cost. Therefore, solar applications will be more viable with fluids for which the amount of heat required is small. From Table 2.3, the heat required for a 1 kW power output falls in the range between 12-20 kW and 6-19 kW for the low-temperature and medium-temperature solar ORC system, respectively. In addition, at a higher turbine inlet temperature, the system requires less heat input, which can be explained by figures 2.7 and 2.8. High temperature saturated vapors reduce the heat input. When designing a solar ORC, one could choose between a system with a larger collector area-low temperature and a system with a small collector area-high operating temperature depending on the application.

The area of solar collector in ORC system is determined by the heat input, net work done and solar insolation of the location for installation. A high turbine inlet temperature requires less area for the solar collector for the system. For a 1 kW power output, the required area of the collector is in the range, 23-34 m^2, for the low-temperature, whereas for the medium-temperature solar ORC, the range is between 14-17 m^2, as listed in Table 2.3. The collector area also constitutes a major part of the system cost. The lowest required area of solar collector is in order RC318, R600a, and R717 in the low-temperature ORC. The lowest heat input fluid is R123 followed by R245fa, methanol, ethanol, R141b, and R11 for the medium-temperature solar ORC system.

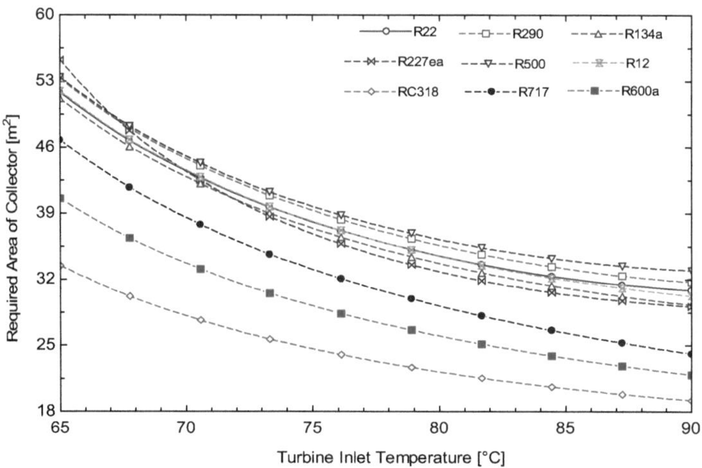

Figure 2.7 Variation of the area of collector for low-temperature solar ORC working fluids with the turbine inlet temperature at a condensing temperature of 45°C

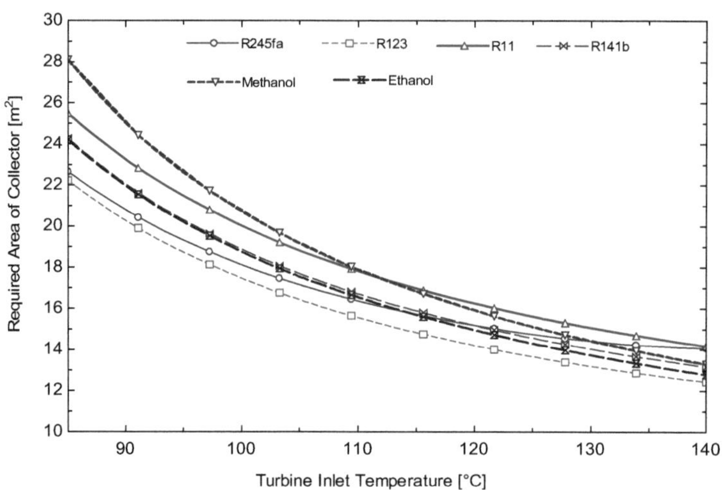

Figure 2.8 Variation of the area of collector for medium-temperature solar ORC working fluids with the turbine inlet temperature at a condensing temperature of 45°C

2.3 Hot water production as co-generation

For production of hot water, the area of collector plays vital role. The other factors include solar radiation, ambient temperature and solar collector efficiency. Hot water can be provided with the condensing temperature from condenser of ORC system. If there is plenty of water in the located site of ORC then this concept of co-generation is widely applied. In the present study, the obtained hot water temperature is 45 °C. The Table 2.4 estimates the average daily hot water production in different months of the year from the required area of solar collector for working fluids. From this, it can be known that larger collector area and higher solar insolation produces more hot water. In the month of April, May, June, July and August, the production is maximal. The highest production is for R500 which has collector area of 30.05 m^2 and produces 2898 Liters per day of hot water followed by R290, R22, R134a in low-temperature whereas in medium-temperature solar ORC the highest production is for R11 which is 1382 liters per day followed by R245fa, ethanol , methanol and R123. The hot water so produced is stored in the insulated storage tank. The application of hot water in rural areas of developing countries are washing utensils, bathing, washing clothes and can be used in small community based hospitals, health posts and clinics.

Table 2.4 Hot water production for different working fluids in a year in Busan, South Korea

Meteorological conditions for Busan, South Korea			Yearly hot water production for various working fluids with their required area of collector														
			Heat source at 90 °C									Heat source at 125 °C					
Month	Solar insolation (W/m^2)	Ambient temp.(°C)	R22	R290	R134a	R227ea	R500	R12	RC318	R717	R600a	R245fa	R123	R11	R141b	Methanol	Ethanol
Jan.	483	3	1336	1378	1292	1116	1415	1327	860	1094	984	631	590	675	631	608	621
Feb.	592	4	1684	1738	1630	1408	1785	1674	1085	1380	1241	796	744	851	795	766	784
March	703	8	2196	2267	2125	1835	2328	2183	1414	1799	1618	1038	971	1110	1037	999	1022
April	877	13	3172	3273	3069	2651	3361	3152	2042	2598	2337	1499	1402	1603	1498	1443	1476
May	928	17	3851	3975	3727	3219	4082	3827	2480	3155	2837	1820	1702	1947	1818	1753	1792
June	842	21	3979	4106	3850	3325	4216	3954	2562	3259	2931	1880	1758	2011	1878	1810	1851
July	733	24	3958	4085	3830	3308	4195	3934	2549	3242	2916	1870	1749	2001	1869	1801	1842
Aug.	718	25	4123	4256	3990	3446	4370	4098	2655	3378	3038	1948	1822	2084	1947	1876	1919
Sept.	620	22	3115	3215	3014	2603	3301	3096	2006	2552	2295	1472	1377	1575	1471	1418	1450
Oct.	593	17	2467	2546	2387	2062	2614	2452	1589	2021	1817	1166	1090	1247	1165	1123	1148
Nov.	472	12	1638	1690	1585	1369	1736	1628	1055	1342	1207	774	724	828	773	745	762
Dec.	440	6	1300	1342	1258	1087	1378	1292	837	1065	958	614	575	657	614	592	605
Daily average hot water production (Liter/ day)			2735	2823	2646	2286	2898	2718	1761	2240	2015	1292	1209	1382	1291	1244	1273

2.4 Shortlist of working fluids for Solar ORC system

According to the operating conditions and various parameters for estimation of 1 kW solar power output ORC, it can be established that most of the working fluids are not accepted though the simulated results yield high thermal efficiencies and pressure ratio. This is because of safety, toxicity, flammability and environmental characteristics of working fluids. In Table 2.5, the decision criteria for working fluids have been recommended. For the value of the parameter preferring the working fluid, it is marked as "√", and in contradictory it is "x". According the decision criteria Table 2.5, the following working fluids are rejected because of following reasons:

High ODP (R500, R123, R11, R141b), high GWP (R22, R227ea, RC318), low safety (R290), high pressure ratio (Methanol, Ethanol), large collector area (R12), low hot water production (R717, R600a).

The only two working fluids (R134a and R245fa) are finally acceptable for operating in the low-temperature and medium-temperature solar ORC system whose heat source temperature are 90 °C and 125 °C respectively.

Table 2.5 Decision criteria table for selection of working fluids

Working Fluid	P_{max}(kPa)	P_{min} (kPa)	PR	η_{exg} (%)	η_{th} (%)	η_{SPC} (%)	A_c (m^2)	Safety Factor	ODP	GWP	Hot water production(L/day)	Decision
R22	x	x	√	√	√	√	x	x	x	x	√	Rejected
R290	x	x	√	√	√	√	x	x	√	√	√	Rejected
R134a	√	√	√	√	√	√	x	√	√	√	√	**Accepted**
R227ea	√	√	√	√	√	√	√	√	√	x	√	Rejected
R500	√	√	√	√	√	√	x	√	x	x	√	Rejected
R12	√	√	√	√	√	√	x	√	√	x	√	Rejected
RC318	x	√	x	√	√	√	√	√	√	x	x	Rejected
R717	x	x	√	√	√	√	√	x	√	√	x	Rejected
R600a	√	√	√	√	√	√	√	√	√	√	x	Rejected
R245fa	√	√	√	√	√	√	√	√	√	√	√	**Accepted**
R123	√	√	√	√	√	√	√	√	x	x	x	Rejected
R11	√	√	√	√	√	√	√	√	x	x	√	Rejected

R141b	√	√	√	√	√	√	√	√	x	√	√	Rejected
Methanol	√	x	x	√	√	√	√	N.a	N.a	N.a	x	Rejected
Ethanol	√	x	x	√	√	√	√	N.a	N.a	N.a	√	Rejected

2.5 Conclusions

A 1 kW solar organic Rankine cycle system has been studied and thermodynamically modeled with 15 selected organic working fluids for low-temperature and medium-temperature on the basis of its critical temperature which ranges from 150 °C to 240 °C. The analysis was carried on comparing various parameters such as exergy, thermal, solar power cycle efficiency in addition to pressure ratio, mass flow rate, heat input, turbine inlet volume flow rate, and area of collector and hot water production. The maximum exergy efficiency is 48.19% and 50.03% for RC318 and R123 respectively. The maximum turbine inlet pressure is 4037 kPa for R22 and R245fa is 1929 kPa. The maximum pressure accounts for RC318 and ethanol. The minimum heat inputs required are for RC318 and R123. Ideally high efficiency, high pressure ratio, high hot water production, low mass flow rate and small area of collector are appropriate. But due to other factors such as flammability, toxicity, and environmental conditions many working fluids are rejected. So the most suitable working fluids recommended are R134a and R245fa for low-temperature and medium-temperature solar ORC respectively. These recommended working fluids requires appropriate collector area size and the production of hot water is reasonable. This Solar ORC can be mostly suitable for remote places of developing countries that lack electricity. Hence this technology is expected to be used for, small distributed power generations systems and producing hot water with the same unit.

References

1. Gao, W., Li, H., Xu, G., & Quan, Y. (2015). Working fluid selection and preliminary design of a solar organic Rankine cycle system. Environmental Progress & Sustainable Energy, 34(2), 619-626.
2. Mavrou, P., Papadopoulos, A. I., Stijepovic, M. Z., Seferlis, P., Linke, P., & Voutetakis, S. (2015). Novel and conventional working fluid mixtures for solar Rankine cycles: Performance assessment and multi-criteria selection. Applied Thermal Engineering, 75, 384-396.
3. A review of working fluid and expander selections for organic Rankine cycle. Renewable and Sustainable Energy Reviews, 24, 325-342.
4. Rayegan, R., & Tao, Y. X. (2011). A procedure to select working fluids for Solar Organic Rankine Cycles (ORCs). Renewable Energy, 36(2), 659-670.
5. Gao, H., Liu, C., He, C., Xu, X., Wu, S., & Li, Y. (2012). Performance analysis and working fluid selection of a supercritical organic Rankine cycle for low grade waste heat recovery. Energies, 5(9), 3233-3247.
6. Saleh, B., Koglbauer, G., Wendland, M., & Fischer, J. (2007). Working fluids for low-temperature organic Rankine cycles. Energy, 32(7), 1210-1221.
7. Maizza, V., & Maizza, A. (2001). Unconventional working fluids in organic Rankine-cycles for waste energy recovery systems. Applied thermal engineering, 21(3), 381-390.
8. Tchanche, B. F., Papadakis, G., Lambrinos, G., & Frangoudakis, A. (2009). Fluid selection for a low-temperature solar organic Rankine cycle. Applied Thermal Engineering, 29(11), 2468-2476.
9. Baral, S., & Kim, K. C. (2014). Thermodynamic modeling of the solar organic Rankine cycle with selected organic working fluids for cogeneration. Distributed Generation & Alternative Energy Journal, 29(3), 7-34.
10. Engineering Equation Solver (EES) .http://www.fchart.com/ees

11. El-Emam, R. S., & Dincer, I. (2013). Exergy and exergoeconomic analyses and optimization of geothermal organic Rankine cycle. Applied Thermal Engineering, 59(1), 435-444.

12. Korea Meteorological Administration (2015). Monthly and seasonal climate summary. Available online: http://www. web.kma.go.kr

Chapter Three: Experimental analysis small-scale organic Rankine cycle

3.1 Description of ORC Prototype

A 1 kW prototype ORC system was designed, built and installed on a laboratory test bench. The prototype included a commercial oil free scroll expander that adopted a magnetic coupling (E15H22N4.25, Air Squared, Broomfield, CO, USA), plate type evaporator (CB60-14H-F, Alfa Laval, Lund, Sweden), plate type condenser (CB76-50E, Lund, Sweden), working fluid variable speed circulation pump (2SF29ELS, Cat Pumps, Minneapolis, MN, USA), and a receiver. In order to perform the experiment for production of the electrical power, different components are assembled together. The following figures 3.1-3.5 below are the components needed for the experiment.

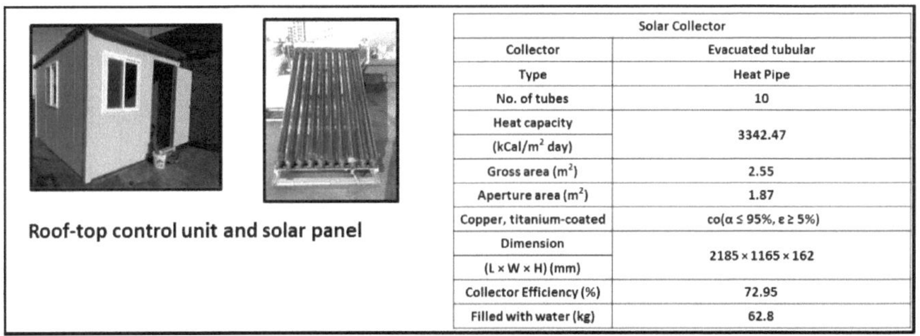

Solar Collector	
Collector	Evacuated tubular
Type	Heat Pipe
No. of tubes	10
Heat capacity (kCal/m^2 day)	3342.47
Gross area (m^2)	2.55
Aperture area (m^2)	1.87
Copper, titanium-coated	co($\alpha \le 95\%$, $\varepsilon \ge 5\%$)
Dimension (L × W × H) (mm)	2185 × 1165 × 162
Collector Efficiency (%)	72.95
Filled with water (kg)	62.8

Figure 3.1 Photograph showing the solar panel and the building

Figure 3.2 Photograph of brazed plate type heat exchangers

Figure 3.3 Photograph of different types of pumps

Scroll expander (13.8 bar/ 175°C)
Model :E15H022A-SH
Volume ratio :3.5
Maximum speed :3600 RPM
Airsquared

Figure 3.4 Photograph of scroll expander coupled with servomotor

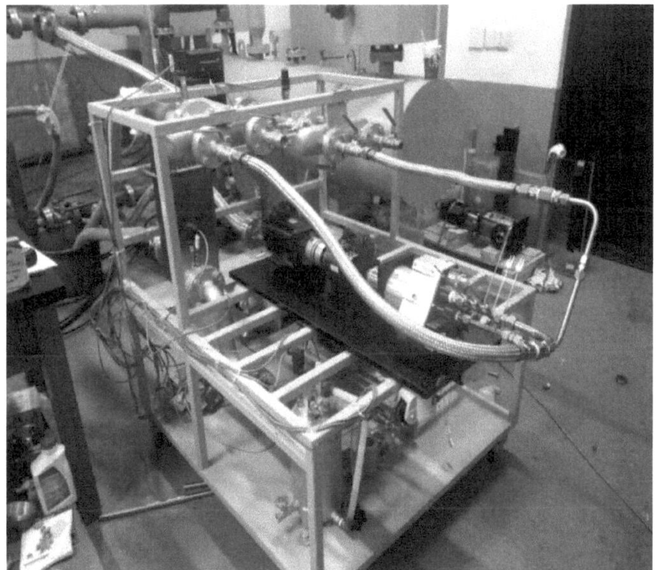

Figure 3.5 Assembled experimental setup in laboratory

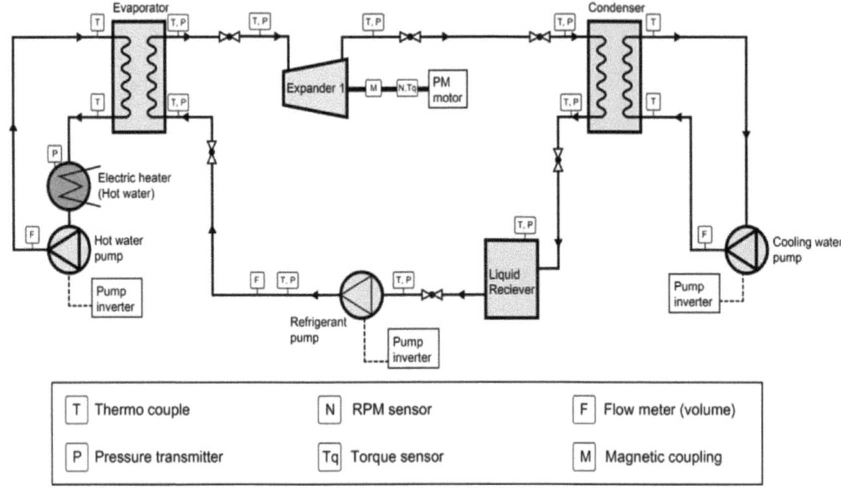

Figure 3.6 Schematic of experimental setup with measuring devices

To obtain the experimental data, the electric thermal heater was allowed to heat water up to 121 °C, and then system was run for approximately 1 hour under different conditions for different cases. The experiments were carried out at different rotating speeds and inlet pressures; the investigated speed range of the expander was 2400–3600 RPM and the inlet pressure range was 10–13 bar. The steady state condition was maintained for 30 min to gather data in each case. The six different cases assessed were as follows: (1) 2400 RPM in 10 bar and 13 bar; (2) 3000 RPM in 10 bar and 13 bar; and (3) 3600 RPM in 10 bar and 13 bar. The schematic of experimental setup is shown in figure 3.6. The thermodynamic properties at different states for particular experiment can be shown in Table 3.1.

Table 3.1 Properties at each state for the ORC system

State No.	Fluid	Phase	Temperature (°C)	Pressure (kPa)	Density (kg/m^3)	Enthalpy (kJ/kg)	Entropy (J/kg·K)
a	R245fa	Dead State	25	101	5.59	424.63	1.705
b	Water	Dead State	25	101	997.05	104.92	0.370
1	Hot Water	Compressed Liquid	121	205.05	942.3	508.06	1.540
2	Hot Water	Compressed Liquid	113	158.44	948.64	474.12	1.460
3	R245fa	Saturated Vapor	89.54	1000	56.4	468.04	1.785
4	R245fa	Vapor	53	210	11.07	446.88	1.810
5	R245fa	Saturated Liquid	35	210	11.92	245.5	1.150
6	R245fa	Compressed Liquid	36	1000	1310.9	247.35	1.160
7	Cold Water	Compressed Liquid	26	195	996.83	109.19	0.380
8	Cold Water	Compressed Liquid	32	195	995.07	134.27	0.470

3.2 Results and discussion

To identify the optimal operating parameters for a small scale ORC system, a test bench was realized to test the expander for different inlet pressures, working fluid mass flow rates and rotational speeds. The aim of the experiment was to determine the highest energy performance that can be used in rural areas of un-electrified countries with minimum maintenance. The highest maximum operating pressure, isentropic efficiency and power output of the scroll expander are the main parameters that need to be established before installing this small-scale ORC system in a particular location of a developing country. A different series of experimental tests was performed to measure the previously described parameters.

Figures 3.7 and 3.8 show the expander power output and thermal efficiencies for different sets of experiments (2400 RPM: 10/13 bar; 3000 RPM: 10/13 bar; 3600 RPM: 10/13 bar). The experimental results indicate that the maximum power output (1.4 kW) can be obtained from the 3600 RPM and 13 bar inlet pressure. The corresponding thermal efficiency was 8.55% at the rotational speed of the expander.

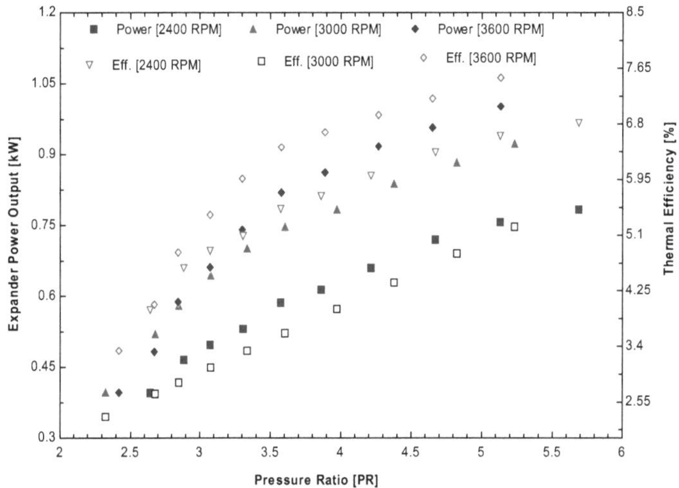

Figure 3.7 Performance parameters: Scroll expander power output and thermal efficiency as a function of pressure ratio at different rotational speeds (10 bar)

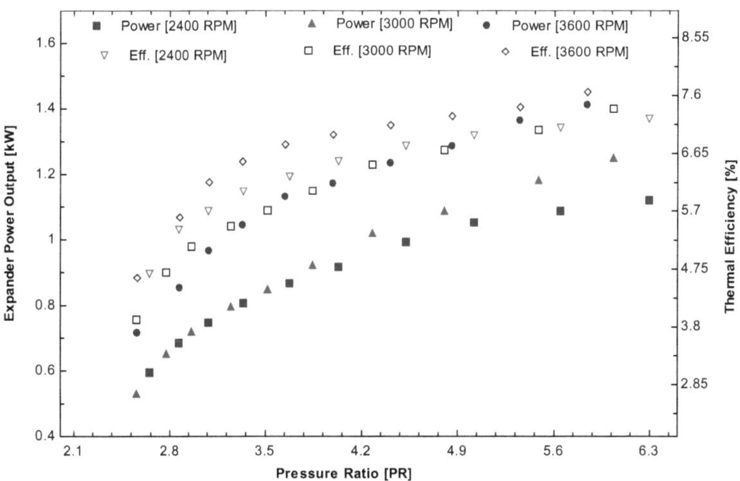

Figure 3.8 Performance parameters: Scroll expander power output and thermal efficiency as a function of pressure ratio at different rotational speeds (13 bar)

The trends for all cases were the same with different rotational speeds of the expander, which demonstrates a higher pressure ratio yield higher power and thermal efficiency. From the experiment, the thermal efficiency of the ORC system increases slightly with increasing rotational speed of the expander but the increase in power output was quite high compared to the expander running from 2400 to 3600 RPM with the same inlet pressure.

It is crucial to investigate the isentropic efficiency of the expander for this small-scale ORC system. This isentropic efficiency of scroll expander helps in determining the total power output of the ORC system. The isentropic efficiency of expander needs to be determined. Figure 3.9 shows that the isentropic efficiency of the scroll expander reached a maximum of 70% when the ORC system was operated at 3600 RPM with an inlet pressure of 13 bar.

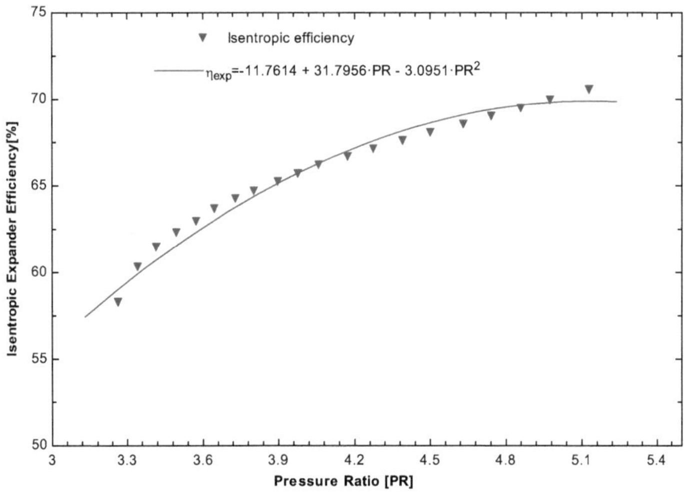

Figure 3.9 Performance parameters: Scroll expander isentropic efficiency as a function of the pressure ratio (2400 RPM/10 bar)

A series of another tests were conducted in the laboratory with different heat source temperature to find the performance, the maximum expander output and thermal efficiency. The low heat source temperature was set at 90 °C, 100 °C and 110 °C and operated with the maximum rotational speed of scroll expander (3600 RPM). Figure 3.10 illustrates the variations of expander output power and thermal efficiency with respect to pressure ratio. It was seen that ORC system operating at 90 °C, 100 °C and 110 °C produced maximum power outputs of 0.67 kW, 0.78 kW and 0.87 kW, respectively. The maximum thermal efficiency was found to be 6.5%, 7.1% and 7.5% respectively at the given heat source temperature and pressure ratio. The variation in power output and thermal efficiency is due to low evaporation temperature and pressure. The enthalpy drop across expander decreases and the mass flow rate decreases with decrease in low heat source operating ORC system. If there is no enough sunshine hour (winter season) or partly cloudy climatic conditions, the solar ORC system could be operated in these low heat source temperatures. From the safety and economic efficiency lower pressure and lower mass flow rate ORC system is more suitable.

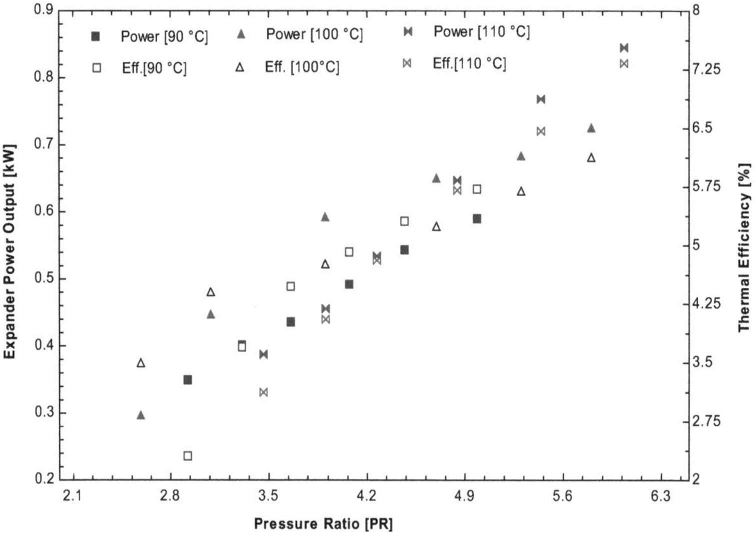

Figure 3.10 Performance parameters: Scroll expander power output and thermal efficiency as a function of pressure ratio at different heat source temperature (3600 RPM)

The success of any technology aimed to install in the rural areas of developing countries depends not only on the thermal efficiency of the system but also on affordability and reliability of the ORC system. The performance of the solar ORC system depends highly on the incident solar isolation, which varies with the geographical position, the time of day and the type of solar collector used. The size of the solar thermal collectors must be well matched in order to obtained highest yield of mechanical and electrical power for the lowest possible cost. The size of the solar collector area can be obtained from the heat input gained during the ORC system. The working fluid mass flow determines the heat energy contained in the ORC system.

Figure 3.11 illustrates amount of heat input in the ORC system to produce the mechanical power with respect to mass flow rate of working fluid. For the ORC system operating at 13 bar (3600 RPM), the maximum heat input or heat required to obtain the power output of 1.4 kW need 14.5 kW of heat supply with the working fluid flow rate 0.07 kg/s. On other hand, 10.5 kW heat energy is required to produce 0.98 kW power output with the working flow rate of 0.045 kg/s (10 bar/2400 RPM). It is seen that working fluid mass flow rate depends on the expander's rotational speed. Higher the operating pressure and rotational speed, the higher is the mass flow rate and heat input. From the literature [1], it is suggested that the power output of 1 kW requires 25 m^2 of solar collector area.

Similarly, from the figure 3.12, the heat input/heat required to produce 0.67 kW of power output requires 8.8 kW of heat energy when the ORC system operating at 90 °C. The working fluid mass flow rate needed is 0.04 kg/s running with the rotational speed of 3600 RPM. In case of 100 °C and 110 °C, the heat inputs are 11.5 kW (0.05 kg/s) and 11.7 kW (0.055 kg/s) respectively. Different types of solar collectors can be used to obtain the heat energy. From this experimental study it is recommended that heat energy input ranges from 9 to 15 kW.

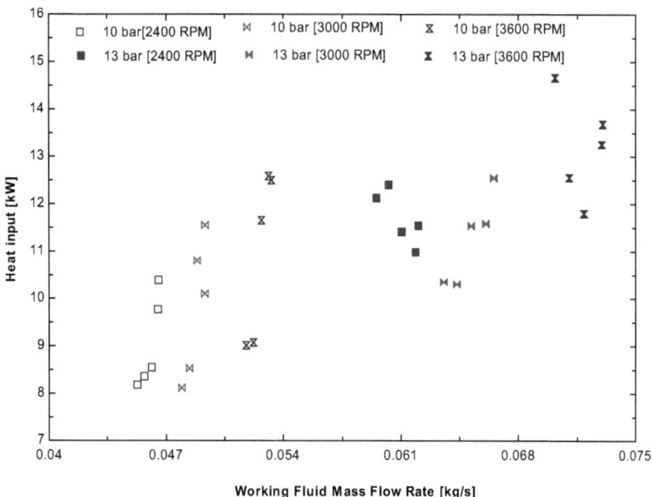

Figure 3.11 Performance parameters: Heat input to the ORC system as a function of working fluid mass flow rate at different the expander inlet pressure and rational speeds

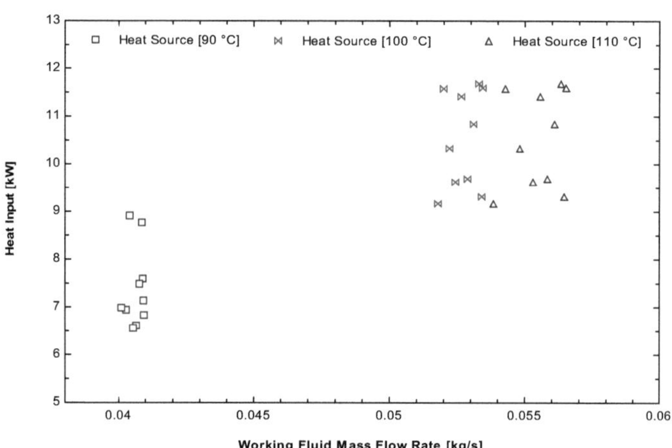

Figure 3.12 Performance parameters: Heat input to the ORC system as a function of working fluid mass flow rate at different heat source temperature (3600 RPM)

3.3 Experimental uncertainty analysis

The physical parameters (data) were directly measured and are as follows:

- The inlet and outlet temperature (°C) for each system's component using K type thermocouples has ±1.1 °C.
- The inlet and outlet pressure (bar) for each system's component using pressure transducer has ±0.044% full scale.
- The working fluid mass flow is measured by turbine type flow meter and has uncertainty range of ±0.4 (L/min) full scale.
- According to the theory of propagation of error[2], the uncertainty estimated errors on the system's efficiency is determined by using the following equations:

$$\frac{\Delta\eta}{\eta}=\frac{\Delta\dot{W}_t}{\dot{W}_t}+\frac{\Delta\dot{W}_p}{\dot{W}_p}+\frac{\Delta\dot{W}_{hw}}{\dot{W}_{hw}} \tag{1}$$

$$\frac{\Delta W}{W}=\frac{\Delta\dot{M}_f}{\dot{M}_f}+\frac{T_{in}+T_{out}}{\left|T_{in}-T_{out}\right|}.\frac{\Delta T}{T} \tag{2}$$

From the figures 3.13 and 3.14, the analysis showed that expander power out has ± 0.042 kW and thermal efficiency has ± 0.13 % change in uncertainty.

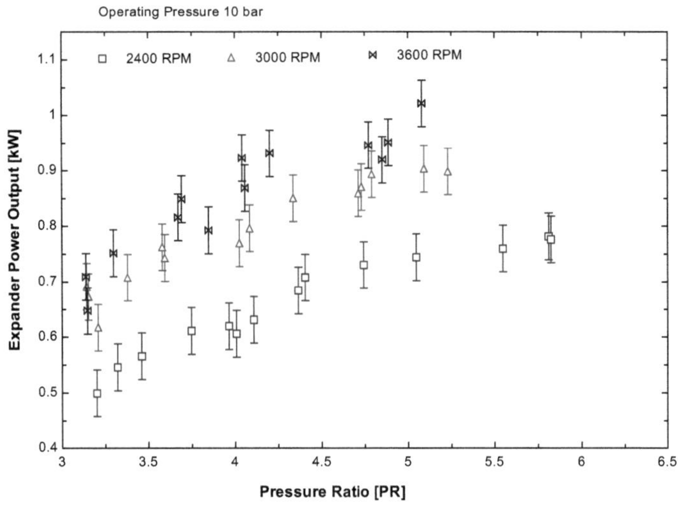

Figure 3.13 Function of pressure ratio as expander power output with error analysis (10 bar)

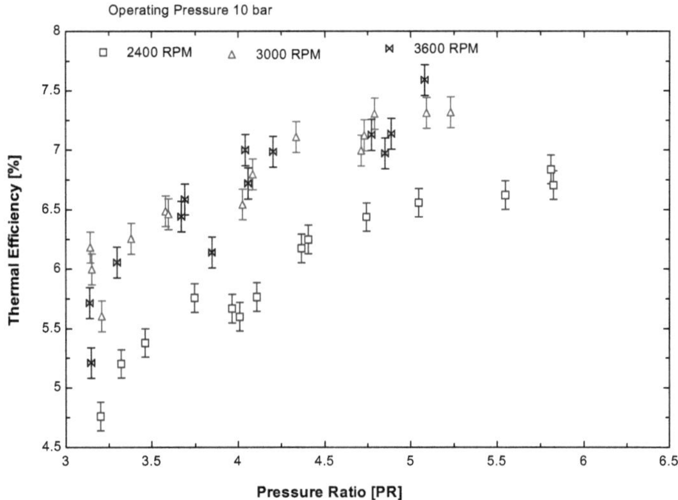

Figure 3.14 Function of pressure ratio as thermal efficiency with error analysis (10 bar)

3.4 Experimental analysis on solar ORC system

The experimental results indicate that the overall solar ORC system can have the maximum of efficiency of 5.2 %. The peak efficiency can be seen when it is nearly in the middle of the day. The solar collector efficiency in this condition is found to be 55 %. The efficiency pattern of the solar ORC system is shown in figure 3.15.

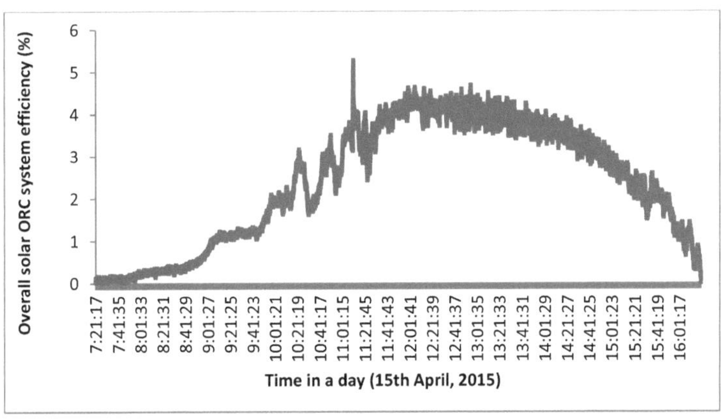

Figure 3.15 Overall solar ORC system efficiency during 15th April, 2015

The solar irradiance values were analyzed further for the hourly solar ORC system power output. The figure 3.16 shows the hourly solar insolation profile for Busan, South Korea [3]. The maximum solar insolation is found to be around 900 W/m^2 during the month of May. Based on this solar radiation, the solar ORC system power output has been simulated. The hourly simulation of solar ORC power output can be seen in figure 3.17. The maximum solar ORC power is in the month of May whereas the lowest is in December. The simulation was carried out with System Advisor Model (SAM) developed by NREL, USA [4].

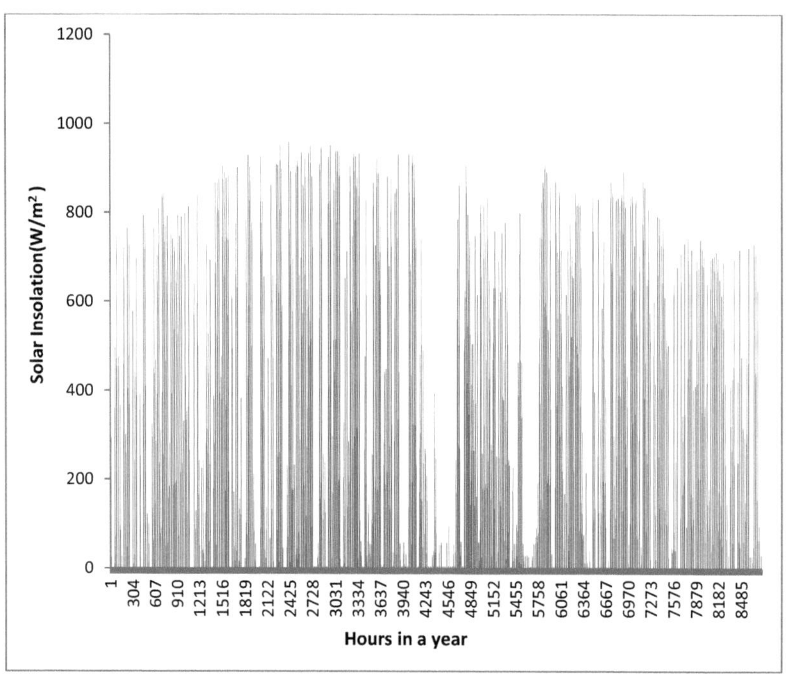

Figure 3.16 Hourly solar insolation for Busan, South Korea

It is seen that the maximum power output to be 1200 W whereas lowest to be 850 W. For the smooth performance of the system, the control unit should be incorporated. This control system allows the optimum power output from the system. If there are cloudy days or low solar radiation due to climatic conditions of the installed location, the system could be automatically shut down and will be turn on when the designed solar ORC is achieved.

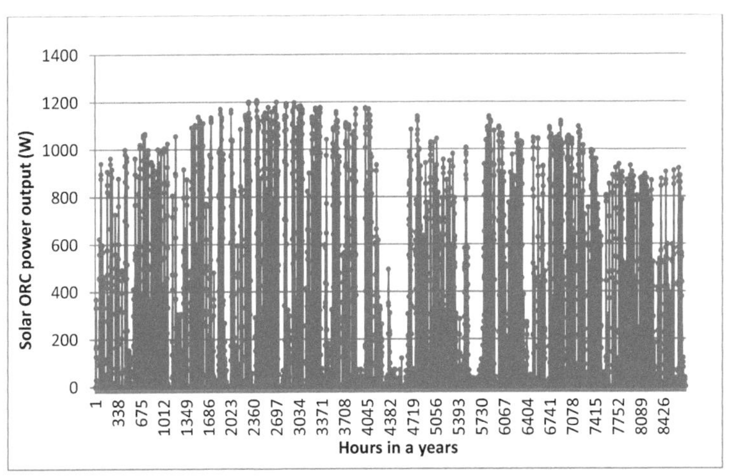

Figure 3.17 Hourly simulation results of solar ORC system during the year

3.5 Conclusions

In this chapter, thermodynamic analysis was carried out to determine the performance, functional and operational parameters of small scale ORC systems. The system's energy and exergy efficiencies were estimated by developing the thermodynamic models. The ORC system was tested in the laboratory aiming to install in the remote areas of developing country. The laboratory tests have shown satisfactory performance over a broad range of conditions including different pressure ratios, rotational speeds of the expander and large variations of the heat source temperature. From the experimental results, the maximum expander power output was 1.4 kW with the expander's rotating speed of 3600 RPM and inlet pressure of 13 bar. The thermal efficiency of the corresponding condition was 8.55% with a maximum pressure ratio of 5.9. No leakage in working fluid from the scroll expander was observed during the experiment. This is because of the magnetic coupling in the scroll. The maximum isentropic efficiency of the expander was found to be 70%. The uncertainity in analysis has also been calculated. The overall solar ORC system has peak efficiency of 5.2 %. The maximum solar power output was found to be 1200 W.The indications thus far confirm the robustness of the ORC system, which could be well adapted in the rural areas of developing countries for electricity production.

References

1. Wang, J., Yan, Z., Zhao, P., & Dai, Y. (2014). Off-design performance analysis of a solar-powered organic Rankine cycle. Energy Conversion and Management, 80, 150-157.
2. Moffat, R. J. (1988). Describing the uncertainties in experimental results. Experimental thermal and fluid science, 1(1), 3-17.
3. Korea Meteorological Administration (2015). Monthly and seasonal climate summary. Available online: http://www. web.kma.go.kr
4. National Renewable energy Laboratory (NREL). https://sam.nrel.gov/ (accessed on 20th May, 2015)

Chapter Four: Thermoeconomic analysis of solar organic Rankine cycle system technology

In this study, the topic of the attribution of a financial value to the thermal energy recovered by the solar ORC system is addressed according to the thermoeconomic approach. Economic analysis of the solar ORC system was carried on by taking into account the purchased components and equipment costs, operation and maintenance (O&M) costs, and the energy input cost. Thermoeconomic analysis is a combination of energy and economic analysis, which provides crucial information that, cannot be obtained through conventional and simple thermodynamic analysis [1-3].

4.1 Estimation of levelized cost of electricity

The thermoeconomic analysis was carried out to determine the cost of electricity production with 120°C solar source temperature. In addition, the payback period and internal rate of return (IRR) of small-scale solar ORC system were calculated. The main expensive components in the solar ORC system are the solar collectors and scroll expander. Table 4.1 lists the cost of different components for estimating the cost of electricity production according to the prototype cost when the solar source temperature is 120 °C, which can produce 1.4 kW of power and annual energy production available equals to 3022.2 kWh. Table 4.2 shows the annual expense of the small-scale solar ORC system. The cost is associated with operation (1%), maintenance (1%) and insurance cost (0.65%) of the cost of electromechanical components [4]. Since this system is simple in construction, robust and has less moving parts, the O&M cost is less. It is concluded that the majority of shared percentage is for thermal energy production system followed by ORC unit and power block. If the costs of these components are reduced, it may be feasible to be adopted in the rural areas of developing countries for electricity generation without huge subsidies.

The prototype cost of solar ORC system is high as compared to other rural electrification technology. It is seen that solar collectors and expander plays an important role in reducing the investment cost of the system.

Table 4.1 Prototype cost of small-scale solar ORC system

Parameters	**Cost($)**	**% of the total cost**	**Economic life (years)**
Thermal energy production unit			
Installation of solar collectors	3100	9.89	
Solar collectors(15 collectors @$ 900)	13500	43.06	20
Collectors pump	450	1.44	20
ORC Unit			
R245fa/water Evaporator	450	1.44	20
R245fa/ Water Condenser	1650	5.26	20
Scroll Expander	3950	12.6	20
R245fa Pump	750	2.39	10
Fluid (R245fa)	150	0.48	
Refrigerant tank and piping	250	0.8	20
Labor cost	200	0.64	
Power Block			
Generator	550	1.75	20
Control systems	300	0.96	20
Others			
Water tank	150	0.48	20
Measuring devices	200	0.64	15
Miscellaneous	150	0.48	
Total Investment cost	**25800**	**100**	

Table 4.2 Annual costs for small-scale solar ORC system

Parameters	Cost($)
Operational cost	247.5
Maintenance cost	247.5
Insurance (Electromechanical equipment)	117.15
Total annual cost	612.15

4.2 Economic analysis of the Solar ORC system

For the economic analysis of the proposed system, it is assumed that the installation of the system is completed within the period of one year. The economic life time of the system is assumed to be 20 years but not all the components' life cycle is so long, for example that of working fluid feed pump and measuring devices. In addition, after 20 years, which is considered to be the life cycle of the system, has no salvage value and is not considered in the present study. The interest rate of 5 % was taken for the analysis. Based on the above assumptions, the annual cash flow rate can be calculated along with present values and annual equivalent cost. The estimation of annual equivalent cost is crucial to find the cost of energy per kWh. The general expression can be written as follows:

$$AEC = \sum_{C=1,N} \frac{IC_C \times (i)}{1-(1+i)^{-n}} \quad (1)$$

where, IC_C represents investment cost of each components which has economic life *(n)* and i denotes the interest rate.

The present value coefficient which correlates a future cash flow with present value is given by:

$$CF_{t=0} = \frac{CF_{t=n}}{(1+i)^n} \tag{2}$$

where, $CF_{t=0}$ is the present value of the future cash flow *(t)*.

The present value of the total cash flows during the economic life cycle of an investment is the net present value (NPV) and is given by:

$$NPV = B_{t=0} - C_{t=0} = \sum_{j=0}^{n} \frac{B_{t=j}}{(1+i)^j} - \sum_{j=0}^{n} \frac{C_{t=j}}{(1+i)^j} \tag{3}$$

where, $B_{t=j}$ is the benefit of the investment in each year and $C_{t=j}$ is the cost of the investment in each year which includes the installation cost in the beginning of its operation.

Benefit-Cost (BC) ratio is an alternative way of expressing the investment criteria and is an index of the ratio of the present value of the benefit cash flows to the present value of the cost cash flows and is given by the following expression:

$$BC = \frac{B_{t=0}}{C_{t=0}} = \frac{\sum_{j=0}^{n} \frac{B_{t=j}}{(1+i)^n}}{\sum_{j=0}^{n} \frac{C_{t=0}}{(1+i)^n}} \tag{4}$$

The payback period (PBP) is the number of years needed for the net present value (NPV) to reach a zero value and is obtained by solving equation (4) with NPV=0 \$, so it is now written as

$$\sum_{j=0}^{PB} \frac{B_{t=j}}{(1+i)^j} = \sum_{j=0}^{PB} \frac{C_{t=j}}{(1+i)^j} \tag{5}$$

The internal rate of return (IRR) is the critical interest rate where the NPV= 0$ and it is always greater than interest rate used for the investment to be economically feasible. IRR can be calculated from the given expression:

$$\sum_{j=0}^{n} \frac{B_{t=j}}{(1+IRR)^{j}} = \sum_{j=0}^{n} \frac{C_{t=j}}{(1+IRR)^{j}} \qquad (6)$$

Equations 1-6 can be found in references [4, 5].For the efficient investment, the benefit to cost ratio (BC) should be larger than 1. The small-scale solar ORC system has BC ratio 1.02 which revealed that the investment is sustainable. From the economic point of view, the Solar ORC system's investment will be profitable after 19 years of installation and operation. This payback period is high and it approaches nearly the life cycle of the system. The main goal of this study is to install the system in rural areas of developing countries for electricity production so it can be treated as social benefit investment. The investment criteria (IRR) in this study have the value 5.01 % is greater than 5%, so it is also favorable investment.

The cost of energy per kWh produced from the system can be calculated from annual equivalent cost divided by annual mechanical energy produced and is equal to 0.68 $/kWh. Figs. 4.1 and 4.2 show annual present value cash flows and net present value (NPV) for system. From the figs., it is seen that the sum of the annual cash flows becomes positive at the end of 19 years of operation which is the payback period. In the 10th and 15th year of the SORC operation it is seen that the gradient of the slope changed, this is due to replacement of components (pump and measuring devices) so the cost of the system is increased.

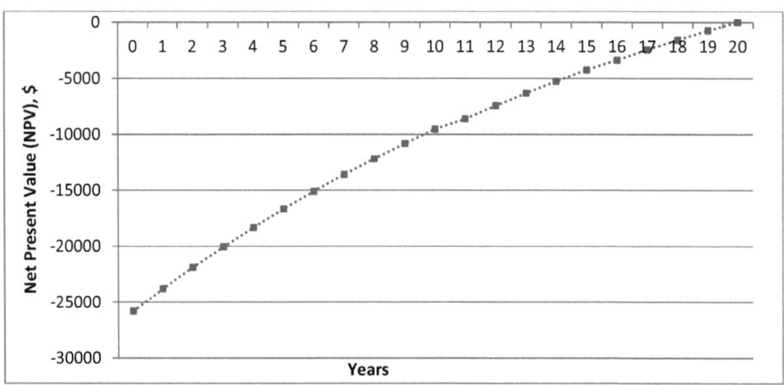

Figure 4.1 Net present values during the life cycle of the system

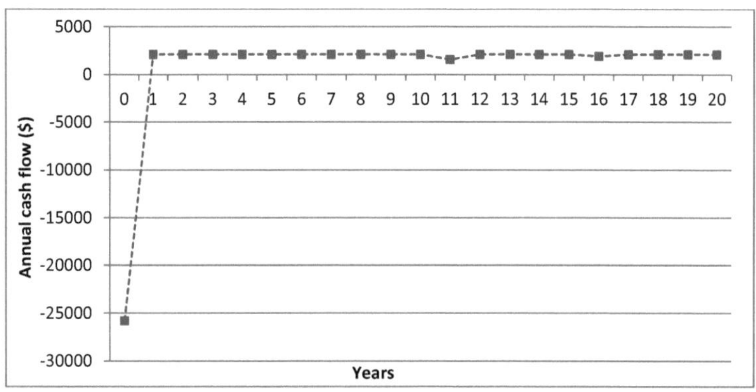

Figure 4.2 Annual cash flows during the life cycle of the system

4.3 Sensitivity analysis

It is one way to glean a sense of all possible outcomes of an investment in order to perform a sensitivity analysis where different values of a certain key variables are tested to see how sensitive investments are to possible change in assumptions. It is the method of evaluating the riskiness of an investment. In the present study, in calculating cash flows, some parameters have more influence on the final result (NPV) than others. The data for different financial scenarios that has been examined for comparison with the standard scenario are presented in Table 4.3. Table 4.4 shows the results of various scenarios for sensitivity analysis.

Table 4.3 Scenarios for examination

Scenario	Economic parameters
1	Standard Scenario
2	20 % increase in investment cost (IC)
3	20% decrease in investment cost (IC)
4	20 % increase in annual benefit (AB)
5	20 % decrease in annual benefit(AB)
6	20 % increase in annual cost (AC)
7	20 % decrease in annual cost (AC)
8	20 % increase in interest rate (IR)
9	20 % decrease in interest rate (IR)

Table 4.4 Results of all scenarios regarding the NPV

Scenario	Investment cost ($)	Annual benefit($)	Annual Cost ($)	Interest Rate (%)	NPV($)
1	25800	2107.85	612.15	5	0
2	30960	2107.85	612.15	5	-5042
3	20640	2107.85	612.15	5	5216
4	25800	2651.85	612.15	5	6834.71
5	25800	1686.28	612.15	5	-6724.17
6	25800	2107.85	734.58	5	-1553.11
7	25800	2107.85	489.72	5	1663.66
8	25800	2107.85	612.15	6	-1991.59
9	25800	2107.85	612.15	4	2382.31

From the analysis, it is observed that the parameters that mostly influence the NPV are the change in the investment cost and annual benefit cost. The least influencing parameters are annual cost and interest rate. Furthermore, Table 4.5 examined all different cases for evaluation of investment. The LCOE differs between 0.86 $/kWh to 0.52 $/kWh, when there is change in net present value from base case (standard scenario). The negative sign in NPV indicates that the investment is not profitable. This can be seen in scenarios 2, 5, 6, and 8. Also the BC ratio less than 1 is not good for investment. The most efficient investment is met under the scenarios 3 and 4 where the payback period is low and has high NPV and IRR.

Table 4.5 Results of all scenarios regarding investment criteria

Scenario	NPV($)	BCR	Payback period (years)	IRR (%)	LCOE ($/kWh)
1	0	1.01	19	5.01	0.68
2	-5154.57	0.87	-	2.94	0.81
3	5165.42	1.18	15	7.82	0.54
4	6834.71	1.2	14	7.82	0.6
5	-5198.42	0.8	-	2.47	0.86
6	-1520.24	0.95	-	4.29	0.64
7	1531.17	1.05	18	6	0.72
8	-1991.6	0.93	-	5.06	0.75
9	2382.31	1.06	17	5.06	0.52

The low payback period of 14 years is estimated according to the tariff 0.6 $/kWh based on scenario 4. The standard scenario has 19 years of payback period with the tariff 0.68 $/kWh. In figure 4.3, the variation of NPV for all scenarios examined illustrated the importance of single variable involved (IC, AB, AC and IR). The greatest dependence of NPV is on the investment cost and annual cost variation for solar ORC. The huge change in NPV is shown by change in investment cost. For the NPV to be positive in comparison with standard scenario, the investment cost, interest rate and annual cost should be increased whereas annual benefit should not be decreased even by certain percentage because the system has long payback period almost equal to the life cycle of system.

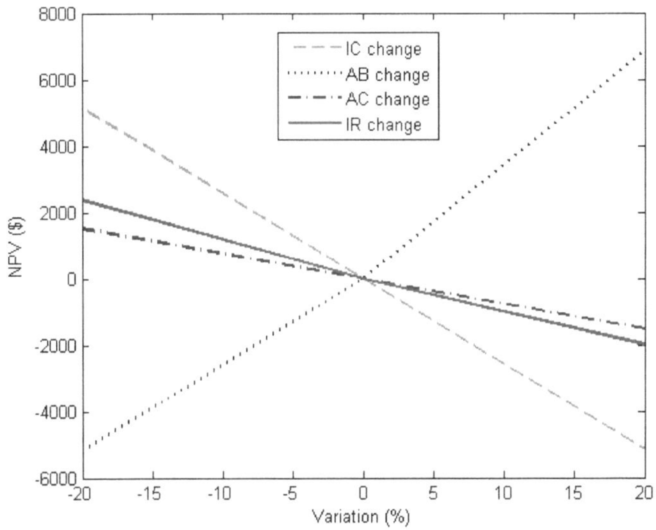

Figure 4.3 Variation of NPV as different variables change (IC, AB, AC, IR)

4.4 Feasibility of investment in South Korea

From the sensitivity analysis, it is seen that the payback period, internal rate of return and net present value highly depends on investment cost, annual benefit, annual cost and interest rate. For the analysis of investment in South Korea, only the interest rate influences the economic parameters. According to the Bank of Korea [6], the interest rate is approximate to 2.2 %. Based on the interest rate, the economic parameters have been estimated. Keeping the total investment cost, annual benefit and annual cost to be same as that of solar ORC, the net present value of the system is found to be $7438.32. Also from the figure 4.4, it is seen that the payback period is 15 years. The internal rate of return and the benefit cost ratio are 3% and 1.2 respectively. The estimation was carried out with the 0.69 $/kWh tariff rate. In the developed country where the interest rate is less, this technology seems feasible for the investment. This technology helps to reduce the GHG emission in the developed country.

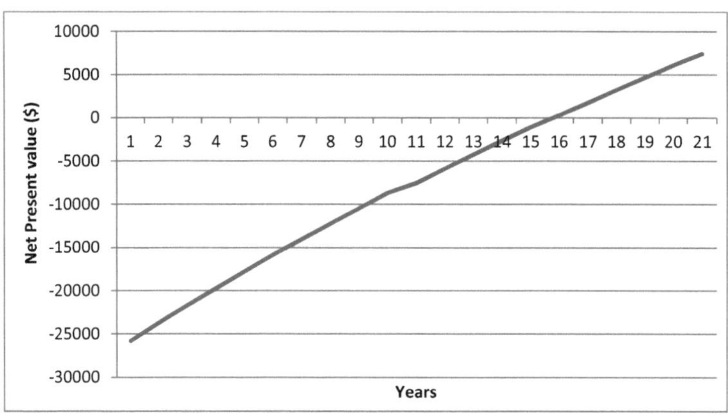

Figure 4.4 Net present values during the life cycle of the system (based on South Korea interest rate)

4.5 Solar ORC based water pumping system

4.5.1 Pumping technology

The rural people in the developing countries are using the diesel generator for irrigation system for the farming. The use of this fossil fuel is making not only environmental pollution but also adds financial burdens for the farmers. In this context, it is necessary to address these issues by developing the concept of pumping system through renewable energy system. The solar ORC technology plays an important role for pumping of water from rivers. This technology also helps rural poor family for electricity generation and water pumping system. The system consists of solar collectors, thermal storage system, ORC unit, motor and pump. Depending on the design and necessity of the system, the battery storage system can also be installed. The use of batteries can play role when the solar intensity is low especially in cloudy, rainy and winter days. However, the battery-less systems can also be proposed for cheaper and simple configuration. The motor has to be chosen according to the power requirements.

4.5.2 Water-pumping power analysis

The power required for pumping water from the river can be determined by the following expression [7]:-

$$P = \rho g Q H \tag{7}$$

where ρ is the density of water (kg/m^3), g is the gravitational acceleration (m/s^2), H is the total dynamic head (m), Q is the volumetric flow rate of water (m^3/s). If the density and gravitational acceleration are kept constant assuming that there is no significant variation, the product QH is directly proportional to the power requirement for pumping water. Hence QH is considered as rate of pumping capacity. The expression (7) can be now expressed as for determining the rate of pumping capacity QH in m^3/s for available power.

$$QH = \frac{P}{\rho g} \tag{8}$$

The volumetric flow rate of required water that can be pumped from river can be calculated by the total head. Therefore for estimating of total pumping capacity for a certain given period of time, the expression can be written as

$$QH \times t = \frac{P \times t}{\rho g} \tag{9}$$

The size of the required pump can be estimated by the following expression which is given by:

$$P = \frac{\rho g Q H}{\eta_p} \tag{10}$$

The pump efficiency (η_p) can be assumed to be 60 %.

4.5.3 Cost estimation of solar ORC water pumping system

For the estimation of cost of pumping of water, the pump (Model: 4W27L2TPJ; Power: 1.5 kW; Head: 166 m; Discharge: 10 LPM; Price: $408) has been taken as from the reference [8]. The hydraulic energy of the system is 99.6 m^4/hr. In the proposed system, the costs of battery storage system and water pump are added with the cost of prototype of solar ORC system. The interest rate is assumed to be same as estimated for solar ORC system which is 5%. The cost of battery is estimated according to the reference [9]. Table 4.6 shows the cost estimation of the solar ORC water pumping system.

Table 4.6 Investment cost of the examined solar ORC water pumping system

	Solar ORC water pumping system without batteries	Solar ORC water pumping system with batteries
Total energy system cost ($)	27645	31690
Annual O&M cost ($)	829.2	950.7
Annual equivalent cost ($)	2218.7	2443.3
Annual pumped water (m^3)	1314	2628

The total cost of technology without storage system is estimated to be \$27645 whereas with the storage system it is \$31690. The battery storage system could be operated for 6 hours daily. The production of water is high due this storage system. Figure 4.5 shows the net present values of the examined solar ORC water pumping system. The net present values are negative until it crosses the 11th year of the installment for both cases. The investment is profitable after 12th year of commencement. Further analysis showed that the cost of pumping water with storage system is \$1.81/m^3 whereas without storage system it is significantly high (\$3.15/m^3). Table 4.7 shows the feasibility indicators in the investment of the solar ORC water pumping system. The internal rates of return, payback periods and benefit cost ratios are almost same in both of the investigated systems which are 5.03 %, 11.7 years and 1.35 respectively.

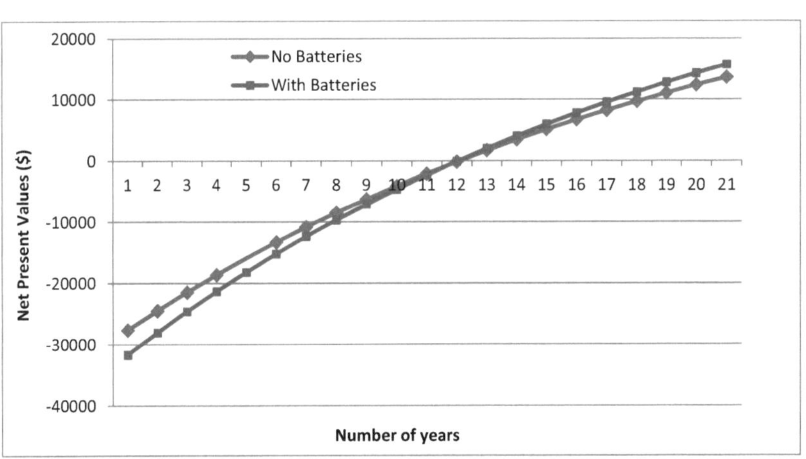

Figure 4.5 Net present values of the examined systems

Table 4.7 Results of investment criteria on solar ORC water pumping system

	Solar ORC water pumping system without batteries	Solar ORC water pumping system with batteries
NPV($)	13614.8	15732.5
IRR (%)	5.03	5.04
PBP(years)	11.7	11.7
BCR	1.35	1.36
Cost of water ($/m^3)	3.15	1.81

4.6 Conclusions

In this chapter of the study, economics and thermoeconomic analyses were conducted to estimate the selling price of produced electricity. The selling price of electricity for solar ORC is 0.68 $/kWh. The payback period for the solar ORC system is 19 years, which is almost equal to life cycle of the system. Since the payback period is very high, this technology is only used in rural areas of developing countries, which lack electricity for lighting homes. The sensitivity analysis was carried out to observe the influence in the NPV of solar ORC system. It is concluded that the reduction in investment and annual costs result in lowering the cost of energy per kWh. Special care should be taken in the estimation of the interest rate together with a realistic estimation of the total cost and the annual cash flow rates for economic analysis because the interest rate has strong influence on the specific cost of the developed system. The prototype cost of the system is very high, so if subsidies are not given the SORC market cannot grow exponentially. The small-scale solar ORC is currently expensive as compared to medium and large scale but if it could be successfully scaled down, the solar ORC could see improved competitive with PV in short term. The mass production of the system components can play an important role in reducing the total investment cost of the system.

Finally, for the investment feasibility in South Korea, the estimation shows that the payback period is 15 years with 3 % internal rate of return. The payback period is less as compared to other case, because of the low of interest rate in South Korea. In other section of this chapter, the concept of solar ORC water pumping system has been introduced. The result shows that the cost of pumping water to be $ 1.81/m^3 and $ 3.15/m^3 for solar ORC water pumping system with storage and without storage system respectively. The cost of pumping is very high in this proposed system. Since the goal of this system to irrigate the land where there is river and lack of electricity for pumping the water. The system could be highly feasible if some subsidies are provided from the government or other agencies for the installation.

References

1. Heberle, F., & Brüggemann, D. (2014). Thermoeconomic analysis of hybrid power plant concepts for geothermal combined heat and power generation. Energies, 7(7), 4482-4497.
2. Al-Sulaiman, F. A., Dincer, I., & Hamdullahpur, F. (2013). Thermoeconomic optimization of three trigeneration systems using organic Rankine cycles: Part I–Formulations. Energy Conversion and Management, 69, 199-208.
3. Kosmadakis, G., Manolakos, D., Kyritsis, S., & Papadakis, G. (2009). Economic assessment of a two-stage solar organic Rankine cycle for reverse osmosis desalination. Renewable energy, 34(6), 1579-1586.
4. Baral, S., Kim, D., Yun, E., & Kim, K. C. (2015). Experimental and Thermoeconomic Analysis of Small-Scale Solar Organic Rankine Cycle (SORC) System. Entropy, 17(4), 2039-2061.
5. Park, C. S. (2004). Fundamentals of engineering economics. Prentice Hall.
6. Bank of Korea. http://www.bok.or.kr/eng/engMain.action (Accessed on: 10th June, 2015)

7. Cimbala, J. M., & Çengel, Y. A. (2008). Essentials of fluid mechanics: fundamentals and applications. McGraw-Hill Higher Education.
8. Crompton pumps residential pumps price. http://www.cromptonpump.com (Accessed on 25th July, 2015)
9. Zakeri, B., & Syri, S. (2015). Electrical energy storage systems: A comparative life cycle cost analysis. Renewable and Sustainable Energy Reviews, 42, 569-596.

Printed by Books on Demand GmbH, Norderstedt / Germany